THE THEORY OF DEATH.

❖ CONTENTS:

- 1) INTRODUCTION
- 2) WHAT IS DEATH?
- 3) HOW THE DEATH IS TAKING PLACE?
- 4) SUPERSTITIONS RELATED TO DEATH.
- 5) EXAMPLES.
- 6) HOW EVERY LIVING ORGANISM CAN SAVE OURSELVES FROM DEATH?
- 7) THE THEORY OF DEATH.

❖ 1. INTRODUCTION

➢ I know you are very eager to know about this book and also about an important process of our life.which is very sad for every human being and also every living organism and also this process this process is the end of one's life.This process of life is known as the 'Death'.

➢ In this book I'm going to tell you about
1)What is Death?
➢ 2)How the death is taking place?
➢ 3)superstition related to Death.
➢ 4)How every living organism can save ourselves from Death?
➢ I am going to explain all this concepts and questions with examples.First I am

starting with my first question and my second content which is:

-

2. What is Death?

➢ Death is the permanent cassation of all biological function that sustain a living organism. Phenomena which commonly bring about death include aging, disease, predation, malnutrition, suicide, homicide, starvation, dehydration and accidents or major trauma resulting in terminal injury.

➢ Death-particularly the death of humans has commonly been considered a sad or unpleasant occasion, due to the affection for the being that has died and the termination of social and familial bonds with the decreased other concerns include fear of death,necrophobia,anxiety,sorrow grief,emotional pain,depression,sympathy,compassion,solitude or saudade. Many cultures and religions have the idea of an afterlife and also hold the idea of reward or judgement and punishment for past sin.

➢ The leading cause of human death in developing countries is infectious disease. The leading causes in developed countries are atherosclerosis(heart disease),cancer and other disease related to obesity and aging. By an extremely wide margin,the largest unifying cause of death in the developed world is biological aging leading to varios complications known as aging-associated diseases.

➢ In developing nation,inferior sanitary conditions and lack of access to modern medical technology makes death from infectious diseases more common in developed countries.one such disease is tuberculosis,a bacterial disease which killed 1.8M people in 2015.Malaria causes fever which killes 1-3M people annaually.AIDS death toll in Africa may reach 90-100M by 2025.

-
- Now I am starting with my second question and my third content which is:

3. How the Death is taking place?

➤ The scout motto is 'Be Prepared' but it's hard to be prepared for death, be it our own or a loved one's. Too much is unknown about what dying feels like or what, if anything happens after you die to ever feel truly ready. However, we do know a bit about the process that occurs in the days and hours leading upto a natural death, and knowning what's going on may be helpful in a loved one's last moments.

➤ During the dying process the body's system shuts down. The dying person has less energy and begins to sleep more and more. The body is conserving the little energy it has and as a result, needs less nourishment and sustenance. In the days or sometimes week) before death, people eat and drink less. They may lose all interest in food and drink and you shouldn't force them to eat. In fact, pushing food or drink on a dying person could cause him or her to choke-at this point, it has become difficult to swallow and the mouth is very dry.

➤ These are the symptoms that a person experiences before dying.

➤ Death is also taking place because of aging, predation, malnutrition, diseases, suicide, homicide, starvation, dehydration and accidents or major trauma resulting in terminal injury.

-
- Now I am starting with my third part and fourth content which is:

4. Superstitions related to Death.

➤ There are many superstitions related to death but the few important superstitions are:

➤ i)Celebrities die in threes:

➤ This one has many modern adherents because it is impossible to disprove.Who qualifies as a celebrity?people die all the time so it's rarely difficult to find somebody even slightly well-known to round out a three some,the origin of the modern superstition might have arisen from an old English folk belief that three funerals tended to occur in rapid succession.why that one arose however,has been lost to posterity.

- ii) Hold your breath while passing through a cemetery:
- Similar to the superstition that we should cover our mouths when yawning to prevent our spirit from leaving our body, holding your breath when passing a cemetery supposedly prevents the spirits of the dead from entering you. (Of course, the real trick is to hold your breath and avoid stepping on any cracks in the sidewalk!)
- iii) Three on a match is Bad Luck.
- iv) Cover the Mirrors in a Home where a death Occurred.
- V) Touch a Button if you see a Hearse.
- The most important superstition and the aim of this book is:

➤ vi)That is every living organism are dying or the death is occurring in living organism because of god.

➤ I am starting to disprove this superstition with some explanation and examples:

➤ In this Book I want to tell every -one that death is not occurring because of God or any other thing in the world.

➤ Death is completely occurring because of ourselves only, because of our fault only death is occurring in every living organisms. As shown in the picture every living organism finishes his life himselves only.

➢ I am starting to disprove the sixth superstition with few examples and the examples are:

❖ 5. Examples:

➢ 1)eg: If you are walking in a road bike or any other vehicle comes and hits you and the accident takes place,if it is a minor accident then you can be saved by doing surgeries but if it is a major accident then it results in your's Death.

➢ When you will clearly think about this example then you can see that your's death is taking place of your's fault only it is not the fault of god or any other thing.

➢ If you had walked carefully in the road then your death will not been occurred and you can be lived for many and many years without facing the death.

➢ My second example is:

- ➢ 2)eg: In some living organism the death is taking place because of disease.
- ➢ If you are getting birth your organs and you are very active at that time no disease can attack you but as the time passes at the age of 30 or after 30 you are getting attacked from disease. This is also one of the superstition related to Death.
- ➢ Not only at the age of 30 or after 30 a person gets attacked by disease it can also attack a small child also. There are instances that a small child getting attacked by disease and that disease results into the death of the child also.
- ➢ As the person failed to maintain his body well, the disease attacks the person and the death is taking place in the person.

➢ From this I want to conclude that Death is not taking place because of age fault or god fault it is completely occurring because of the person's fault only.

➢ My third example is:

➢ 3)eg: In some living organism the death is taking place because of other living organism.

➢ In some cases the first person die because of second person as the second person was the enemy of first person because of various reasons.In this example also the first person die because of his fault only as the first person had done something wrong with the second person that's why the second person had become enemy of first person and kills the first person.

➢ My fourth example is:

➤ In some living organism the death is taking place because of natural occurring such as:

➤ 4)In food chain one organism eats another organism.A seen in the abundance of frog is eaten by snake but if the frog is not seen by the snake then the snake cannot eat the frog i.e.Death of frog is taking place because of his fault only it is not the fault of god or any other living organism.

➤ If the frog is not seen by snake and the frog is maintaining its body well then the frog can live for many and many years without facing the death.

➤ My fifth example is:

➤ 5)eg:In some humans the death is taking place because as they kill themselves only this is also known as suicide.In this death occurring process also the death is taking place because of his fault only as the person kills himself only.In this example the also the god is not involved in death taking process.

➢ By avoiding the suicide the person can be lived for many and many years without facing the death.

➢ From all these examples and explanation

➢ I want to conclude that death is totally occurring because of person's fault only it is not the fault of god or any other living organism and the second thing is by maintaining our body well and being safe with the surrounding every living organism can live for many and many years and lot more then their estimated years. In this book I will not tell say that living organism has no end but every living organism can live more than their estimated years.

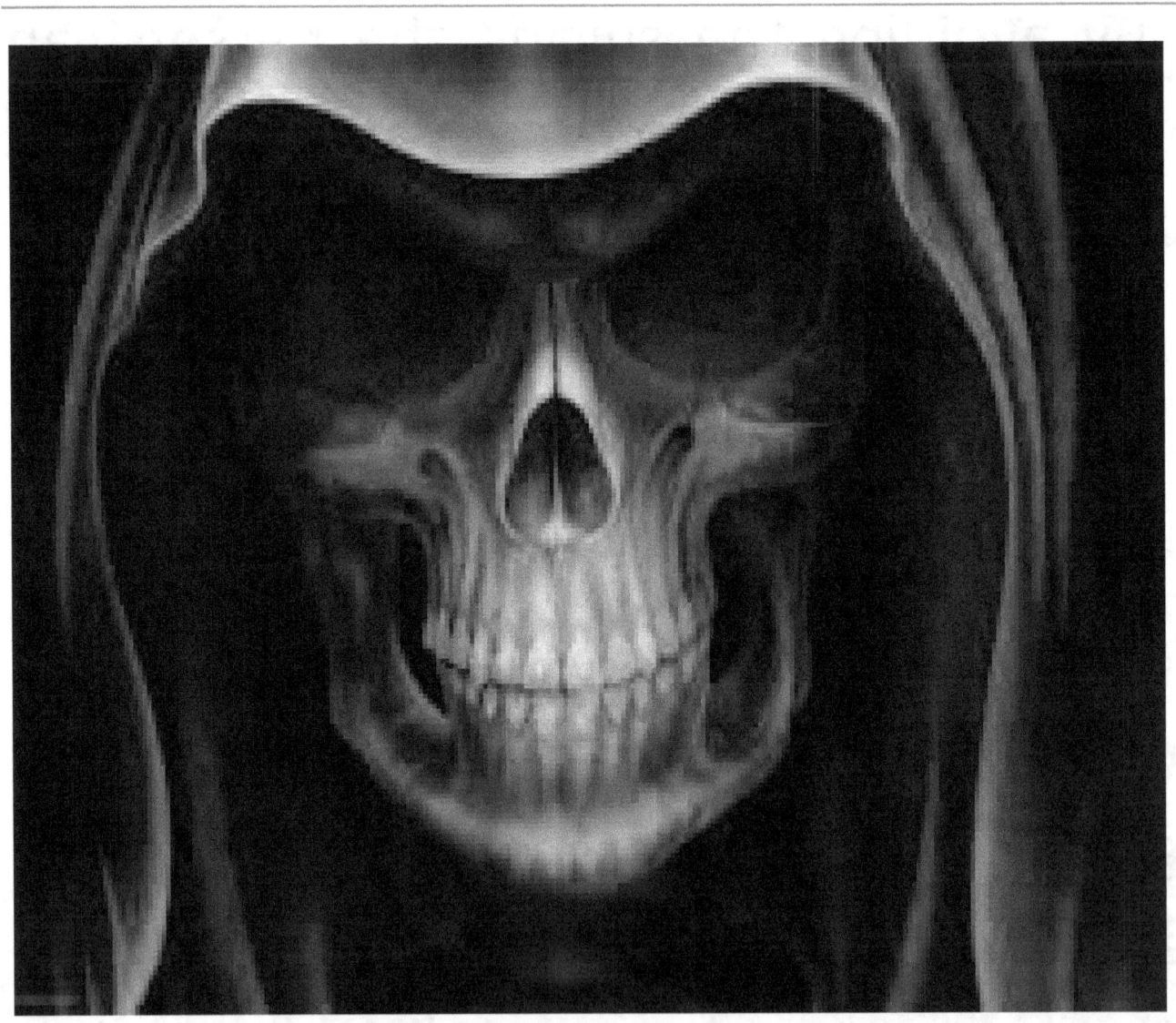

6. How every living organism can save ourselves from Death?

➢ Yes, every living organism can definitively save ourselves from death in the following way:

➢ In the first example we can save ourselves by carefully walking in the road and by following the road rules and traffic rules.

➢ In the second example we can save ourselves by maintaining our body well and by eating very nutritious food and by living in healthy atmosphere and environment.

➢ In this way, in every example of death taking process we can save ourselves by taking care of his life and can live for many and many years without facing the death.

The theory of Death.

- 7.
- From all these explanation and examples I am concluding my theory of death which is:
- The theory of Death include two concepts.
- The first concept is:
- If the living organism maintains its body very well and the living organism is very careful with the surrounding then the living organism can live more than the estimated years without facing the death.

➢ The second concept is:

- If the living organism fails to maintains its body well and the living organism was not able to live carefully with the surrounding or any other thing makes the living organism to go under the process of death.
- In this the death of living organism is completely occurring because of his fault only it is not the fault of god or any other living organism.
- The important point that second concept tells us is:
- The god or any other thing is not involved in the death taking process of living organism it is totally the fault of living organism that the organism goes under the process of death.
- Thank you! For reading my book.

-by Harish Mudaliyar.

STD:X.

THE END.